Favorite Toys

LEGO BRICKS

BY CHRIS BOWMAN

EPIC

BELLWETHER MEDIA • MINNEAPOLIS, MN

Action and adventure collide in EPIC. Plunge into a universe of powerful beasts, hair-raising tales, and high-speed excitement. Astonishing explorations await. Can you handle it?

This is not an official LEGO book. It is not approved by or connected with the LEGO Group.

This edition first published in 2022 by Bellwether Media, Inc.

No part of this publication may be reproduced in whole or in part without written permission of the publisher. For information regarding permission, write to Bellwether Media, Inc., Attention: Permissions Department, 6012 Blue Circle Drive, Minnetonka, MN 55343.

Library of Congress Cataloging-in-Publication Data

Names: Bowman, Chris, 1990- author.
Title: LEGO bricks / by Chris Bowman.
Description: Minneapolis, MN : Bellwether Media, 2022. | Series: Favorite toys | Includes bibliographical references and index. | Audience: Ages 7-12 | Audience: Grades 2-3 | Summary: "Engaging images accompany information about LEGO bricks. The combination of high-interest subject matter and light text is intended for students in grades 2 through 7"– Provided by publisher.
Identifiers: LCCN 2021044250 (print) | LCCN 2021044251 (ebook) | ISBN 9781644876367 (library binding) | ISBN 9781648346477 (ebook)
Subjects: LCSH: LEGO toys–Juvenile literature.
Classification: LCC TS2301.T7 B658 2022 (print) | LCC TS2301.T7 (ebook) | DDC 688.7/2–dc23/eng/20211006
LC record available at https://lccn.loc.gov/2021044250
LC ebook record available at https://lccn.loc.gov/2021044251

Text copyright © 2022 by Bellwether Media, Inc. EPIC and associated logos are trademarks and/or registered trademarks of Bellwether Media, Inc.

Editor: Elizabeth Neuenfeldt   Designer: Josh Brink

Printed in the United States of America, North Mankato, MN.

TABLE OF CONTENTS

Liftoff! 4
The History of LEGO Bricks .. 6
LEGO Bricks Today 16
More Than A Toy 20
Glossary 22
To Learn More 23
Index.................................. 24

Liftoff!

A child grabs his bucket of LEGO bricks. It is time to build a spaceship!

He uses many colorful pieces. Soon, the ship is complete. Anything can be built with LEGO bricks!

The History of LEGO Bricks

LEGO bricks were **invented** by Ole Kirk Christiansen. Ole was a **carpenter** from Denmark. He built furniture and houses. But the **Great Depression** began in 1929. People could no longer afford costly items.

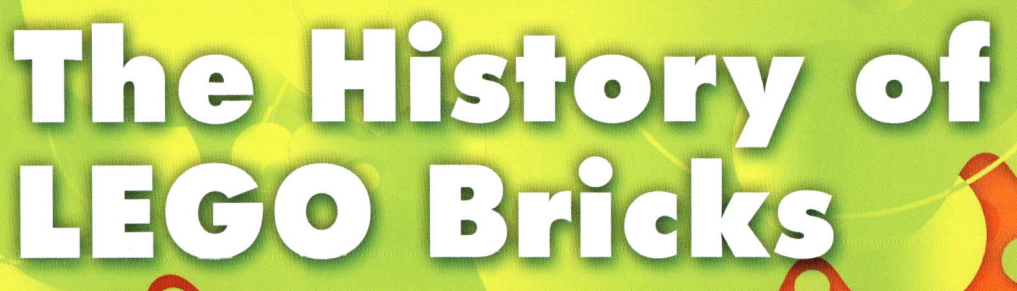

OLE KIRK CHRISTIANSEN

LEGO BRICK BEGINNINGS

Billund, Denmark = 🔴

WOODEN LEGO TRUCK

WOODEN LEGO TRAIN

Ole began making wooden ladders and ironing boards. He also made **model** cars and trains.

People loved the toys he made! Ole soon named his toy company LEGO.

BEHIND THE NAME

LEGO COMES FROM THE DANISH WORDS FOR "PLAY WELL."

LEGO OFFICES IN BILLUND, DENMARK

The LEGO Group started making plastic toys in the late 1940s. The first LEGO bricks were made in 1949. They were called Automatic Binding Bricks.

HOME OF THE BRICK

THE LEGO HOUSE OPENED IN BILLUND, DENMARK, IN 2017. IT IS FILLED WITH ACTIVITIES FOR KIDS AND ADULTS!

AUTOMATIC BINDING BRICKS

LEGO HOUSE IN BILLUND, DENMARK

LEGO TRACTOR

Kids liked the stacking bricks in each set. But other LEGO toys were more popular.

In 1955, LEGO System in Play bricks were created. Bricks from different sets could now be stacked together.

Kids loved these bricks! But the bricks often broke apart. They did not always fit together well.

LEGO SYSTEM IN PLAY SET

LEGO bricks were changed in 1958. Tubes were added to the bottom. Bricks now fit tightly together! In 1974, **figurines** were first included in LEGO sets. **Minifigures** were released four years later.

TUBE

MORE LIFELIKE

MOST MINIFIGURES ARE YELLOW. OVER TIME, LEGO HAS MADE MINIFIGURES WITH MORE NATURAL SKIN TONES.

MINIFIGURES

LEGO BRICK TIMELINE

1934
Ole Kirk Christiansen names his toy company LEGO

1949
LEGO makes its first bricks

1968
The first LEGOLAND park opens in Billund, Denmark

1978
Minifigures start to be included in sets

2014
The LEGO Movie is released

LEGO Bricks Today

Today, there are many kinds of LEGO sets. Sets are often based on movies or TV shows. Some LEGO sets are used to make robots. The MINDSTORMS sets can be **programmed** to talk and move!

HARRY POTTER LEGO

STAR WARS LEGO

LEGO BRICK TYPES

LEGO BOOST

LEGO DOTS

LEGO MINDSTORMS

LEGO DUPLO

LEGO toys are not just for children. Many sets are made for adults. These are often bigger and more **complicated**.

LEGO NES SET

People also collect LEGO sets and minifigures. **Rare** sets and early releases are favorites!

More Than A Toy

The LEGO Movie played in theaters in 2014. It was a hit! More LEGO movies soon followed. LEGO video games and LEGOLAND are also popular. There are many ways to enjoy LEGO bricks!

THE LEGO MOVIE

LEGOLAND PROFILE

What Is It? A chain of LEGO theme parks

When Did It Open? The first opened in 1968 in Billund, Denmark

How Many Are There? There are 10 LEGOLAND parks around the world

Glossary

carpenter—a person who builds or repairs wooden objects

complicated—having more detailed instructions

figurines—statues with human forms

Great Depression—a time in world history when many countries experienced an economic crisis; the Great Depression began in 1929 and lasted through the 1930s.

invented—made for the first time

minifigures—LEGO figurines

model—a small copy of something

programmed—given codes or instructions to do a task

rare—hard to find

To Learn More

AT THE LIBRARY

Dolan, Hannah. *LEGO Minifigure Handbook*. New York, N.Y.: DK Publishing, 2020.

Hugo, Simon. *LEGO: Absolutely Everything You Need to Know*. New York, N.Y.: DK Publishing, 2017.

O'Connor, Jim. *What Is LEGO?* New York, N.Y.: Penguin Workshop, 2020.

ON THE WEB

FACTSURFER

Factsurfer.com gives you a safe, fun way to find more information.

1. Go to www.factsurfer.com.

2. Enter "LEGO bricks" into the search box and click 🔍.

3. Select your book cover to see a list of related content.

Index

adults, 18
Automatic Binding Bricks, 10
beginnings, 7
Billund, Denmark, 7, 9, 10
cars, 8
Christiansen, Ole Kirk, 6, 8, 9
collect, 19
Denmark, 6, 7
figurines, 14
Great Depression, 6
history, 6, 7, 8, 9, 10, 11, 12, 14
kids, 4, 11, 12, 18
LEGO Group, 10
LEGO House, 10
LEGO Movie, The, 20
LEGO System in Play, 12
LEGOLAND, 20, 21
MINDSTORMS, 16
minifigures, 14, 19
movies, 16, 20
name, 9
profile, 21
sets, 11, 12 14, 16, 18, 19
spaceship, 4
timeline, 15
trains, 8
TV shows, 16
types, 17
video games, 20

The images in this book are reproduced through the courtesy of: Levent Konuk, front cover (hero), pp. 22, 23; Domagoj Kovacic, front cover (Land Rover); Ekaterina_Minaeva, front cover (Harry Potter, Emmet Brickowski), back cover (blue car), pp. 16 (Harry Potter), 17 (LEGO Duplo); Shutterstock, front cover (bulldozer); Peter Gudella, front cover (house); cjmacer, front cover (helicopter, LEGO bricks), back cover (robot), p. 4 (LEGO bricks); mini_citizens, front cover (ice cream truck); CTR Photos, front cover (UniKitty); Jeff Bukowski, back cover (X wing); cherry-hai, back cover (spaceship); Divina Epiphania, back cover (Benny's spaceship); Salvador Maniquiz, p. 2 (Chewbacca); droopy76, p. 2 (robot); AlesiaKan, p. 3 (LEGO BOOST robot); focal point, p. 4 (LEGO bucket); AkosHorvath, p. 4 (spaceship); Kuznetsov Dmitriy, p. 5 (child); igra.design, p. 5 (minifigures); ©2021 The LEGO Group., pp. 6, 8 (wooden truck), 10 (Automatic Binding Bricks), 15; seewhatmitchsee, pp. 7 (minifigures, astronaut), 15 (minifigure); Bloomberg/ Getty Images, p. 8 (wooden train); Raimonda Kulikauskiene/ Getty Images, pp. 9, 10; brasletty1, p. 11; Lumella, p. 12; Niels Poulsen/ Alamy, p. 13; TY Lim, p. 14 (minifigures); Visut Apiwatamon, p. 14 (LEGO bricks); Margus Vilbas, p. 15 (The LEGO Movie); aperturesound, p. 16 (Star Wars); Ilsur Nigmatzyanov, p. 17 (LEGO DOTS); Jeff Gilbert/ Alamy, p. 18; Kit Leong, p. 19; Collection Christophel/ Alamy, p. 20; Jeppe Gustafsson, p. 21.